Ben's Weather Friends Explain Fronts, Thunderstorms, and Rainbows

I0025429

Written by Ben & Linda Robinson
Illustrated by Linda Robinson

Hi! I'm Rayne the Raindrop and I'm here to show you how fun weather is.

Do you know what a cold front and a warm front are?
Do you know how thunderstorms form?
No worries if you don't, I have friends that will help me explain.

Here is my friend, Cindy the Cold Front.

Hi! I'm Cindy and I'm here to explain the weather I bring and what I do.

When you think of a cold front, think about a bulldozer.
A cold front, like a bulldozer, pushes the warm air up.

COLD AIR

As the warm air rises into the sky, it gets cooler, forming clouds.

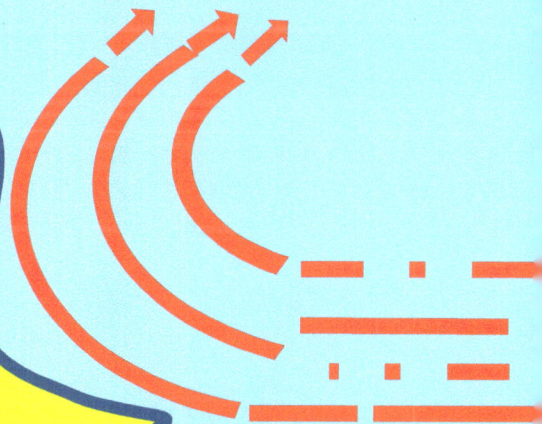

WARM AIR

A cold front can bring lots of friends.
I can bring gusty winds, cold air, and thunderstorms.

Sometimes I bring all my friends, and sometimes I bring just one.
We all love to travel together.

Gus T. Wind

Collet Cold Air

T.J.
Thunderstorm

**Thank you, Cindy.
And now ... William the Warm Front!**

Hi! I'm William the Warm Front.
I'm going to help you understand what I do and the weather I bring.

Think of a warm front as going up a slide backwards.
A warm front is when moist, warm air slides up and over cold air.

Warm fronts can bring clouds, a change in wind direction, and rain.
After a warm front, there will typically be clear skies and warmer temperatures.

Meet the weather friends I may bring!

Claudia Cloud

T.J. Thunderstorm

Rayne Raindrop

Thank you, William.
Now, let's hear from T.J. Thunderstorm!

Hi! I'm T.J., and I'm going to explain the three stages of a thunderstorm.

The first stage is called the developing stage. This is when updrafts (winds going up) push moist warm air up to make a cumulus (ku-mu-lus) cloud.

Developing

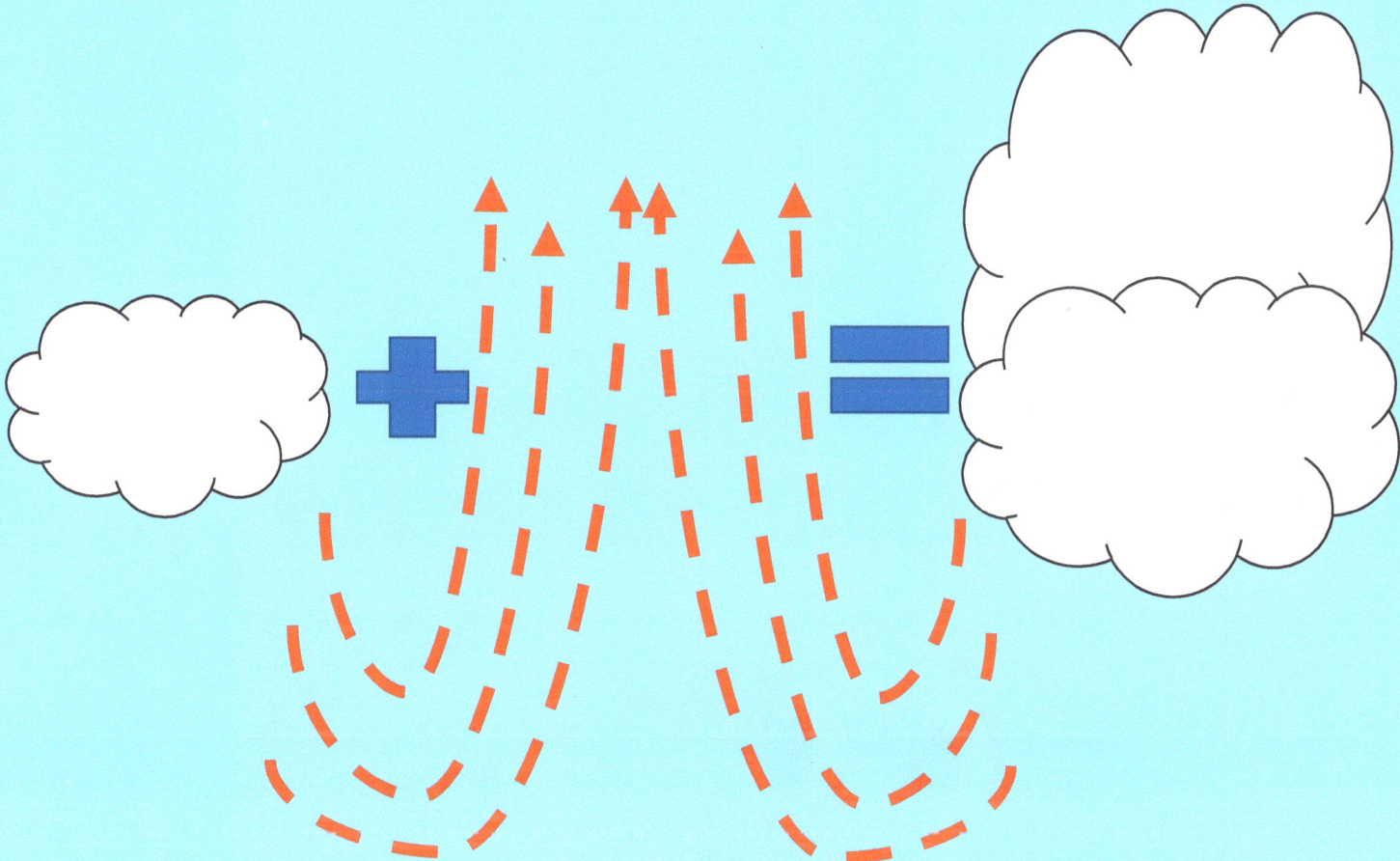

Updrafts

The second stage is the mature (ma-tur) stage.
The cloud continues to grow, collecting more moisture.
The moisture condenses into water droplets.

Mature

Developing

As the cloud grows, the droplets grow.
When the water droplets become heavy, the cloud drops them.
This is a raindrop, like Rayne!

The third stage is the the dissipating stage.
In this stage, the warm air in the cloud begins to cool.

Dissipating

Mature

Developing

The cool air sinks back to the ground and eventually stops the warm moist air from rising. This causes the rain to slow down then stop.

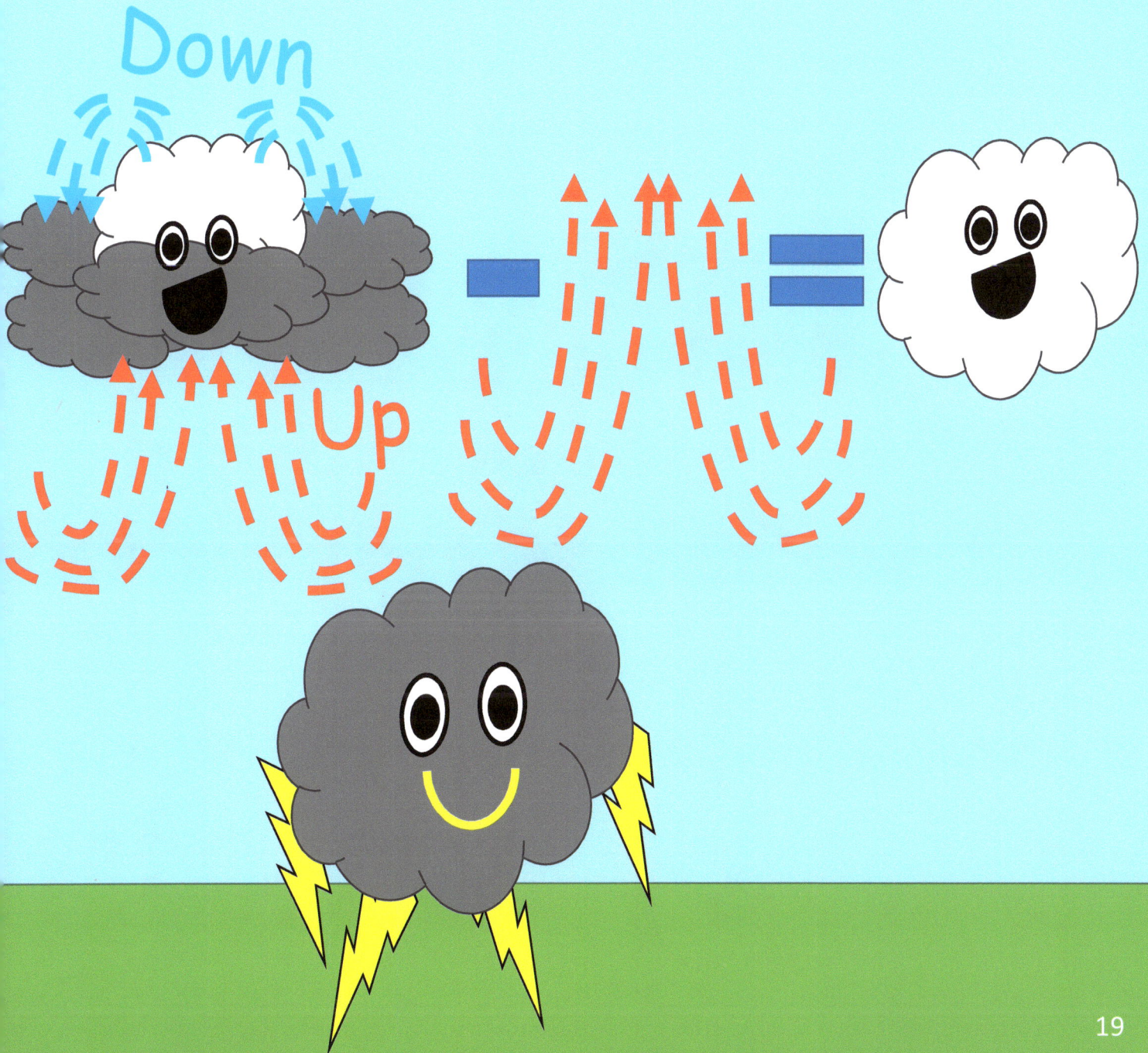

Down

Up

I bet you want to know about lightning and thunder.

In a thunderstorm cloud, two of my friends like to run around and play, Polly Positive Charge and Nick Negative Charge.

Polly and Nick hang around the clouds and on the ground.

After awhile, they begin to miss each other and run to the ground to get more friends. They run so fast they make a lightning bolt.

With running to get friends, this causes the air to heat and expand, then quickly cool and contract. This creates a shock wave known as thunder.

BOOM!

Don't be afraid of thunder, it's just Polly getting Nick to play.
Lightning, however, is dangerous!
It's safest to stay indoors or find shelter during a storm.

Lightning is very HOT and DANGEROUS!

Storms can be scary, but the best thing about storms is that sometimes there is a rainbow at the end. Look for my friend Rene the Rainbow after a storm.

I love to play after the rain!

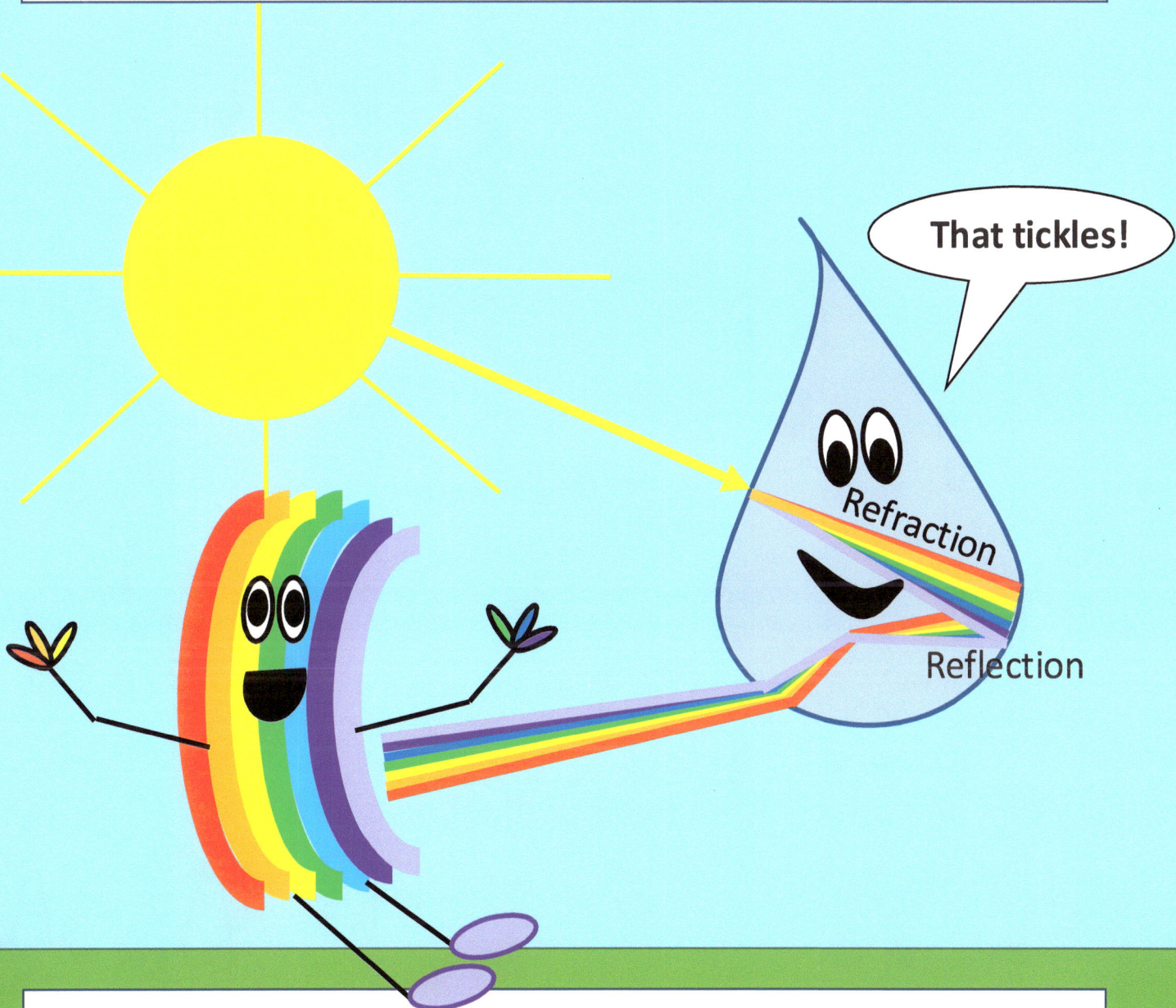

Rainbows form when sunlight enters a raindrop, refracts, then reflects off the inside of the raindrop, separating it into seven different colors: red, orange, yellow, green, blue, indigo and violet.

That tickles!

Refraction

Reflection

In this case, refraction is when the light wave changes direction and reflection is when the light wave comes back.

I hope you understand weather more. Weather is all around you. Learning about weather can be just as much fun as watching it.

Before you go, see how many of my friends you can name.

See you next time it rains!

About the Authors:

In 1991, ten-year-old Ben asked his big sister to help him design a weather book. Between their love of weather and combined imaginations, Randy the Raindrop (renamed Rayne the Raindrop) and friends were born in the original book, *Randy's Weather*. Now, more than thirty years later, they've teamed up again to share their story with all readers.

Linda and Ben Robinson, were both born in Fort Wayne, Indiana. Ben Robinson is a network engineering consultant in Honolulu, Hawaii, with an enthusiasm since childhood for the dynamics of weather and its ever-changing impacts. Linda Robinson is an author, veteran, project manager, and lives in Helotes, Texas with her two children, four dogs, and two cats.

Original illustrations, circa 1991

www.ingramcontent.com/pod-product-compliance
Lightning Source LLC
Chambersburg PA
CBHW060854270326
41934CB00002B/134